QUESTION & WORKBOOK

Working for over 30 YEARS with Cambridge Assessment International Education

Cambridge International AS & A Level

Mathematics

Pure Mathematics 2

Greg Port

HODDER
EDUCATION

Hachette UK's policy is to use papers that are natural, renewable and recyclable products and made from wood grown in well-managed forests and other controlled sources. The logging and manufacturing processes are expected to conform to the environmental regulations of the country of origin.

Orders: please contact Hachette UK Distribution, Hely Hutchinson Centre, Milton Road, Didcot, Oxfordshire, OX11 7HH. Telephone: +44 (0)1235 827827. Email education@hachette.co.uk Lines are open from 9 a.m. to 5 p.m., Monday to Friday. You can also order through our website: www.hoddereducation.com.

ISBN: 978 1 5104 5843 7

© Greg Port 2018

Published by
Hodder Education, an Hachette UK Company
Carmelite House, 50 Victoria Embankment
London EC4Y 0DZ

www.hoddereducation.com

Impression number   10 9 8 7 6 5

Year   2023

Cover photo by Shutterstock/Chatgunner

Illustrations by Integra Software Services

Typeset in Minion Pro Regular 10.5/14 by Integra Software Services Pvt. Ltd., Pondicherry, India

Printed in the UK

A catalogue record for this title is available from the British Library.

Contents

1 Algebra

1.1 Operations with polynomials

1   State which of the expressions below are polynomials. If they are polynomials, state the order of the polynomial.

Expression	Polynomial Yes/No?	Order
$2x - x^2$		
$\dfrac{x}{2} - \dfrac{2}{x}$		
0		
$x^{23} + 2x^{15} + 1$		
$x^3 - 2x^2 + \sqrt{x}$		
$x^2 + \sqrt{2}$		
$5 + x + \pi x^{45}$		
$1 - 3x$		

2   Find the values of the constants A, B, C and D in the identity $x^3 - 4 \equiv (x-1)(Ax^2 + Bx + C) + D$.

OCR MEI Further Concepts for Advanced Mathematics FP1 4755 Paper 01 Q3 June 2007

3   Simplify these expressions as much as possible.

(i) $(3x^3 + 4x^2 - 2x - 1) + (x^3 - 2x^2 + 7x + 1)$

(ii) $(x^3 - 5) \times (3x^3 - 2x + 1)$

4 Use long division to divide these polynomials, giving the quotient and the remainder.

(i) $x + 2 \overline{)x^2 + 3x - 1}$

(ii) $2x + 1 \overline{)2x^4 - x^3 + 15x^2 + 12}$

(iii) $\dfrac{3x^2 + 5x - 1}{x^2 + 1}$

(iv) $\dfrac{x^4 + x^3 - 3x^2 + 2x - 4}{x^2 - x}$

5 In the division $\dfrac{x^3 + 5x + a}{x^2 - b}$ the remainder is 2. Find the values of a and b.

1.2 Solution of polynomial equations

1 (i) Find the remainder when $f(x) = x^3 - 3x^2 - 6x + 8$ is divided by

    (a) $x + 1$ **(b)** $x - 3$

    (c) $2x + 1$ **(d)** $x - 1$

(ii) Using your answers to part **(i)**, fully factorise f(x).

(iii) Hence solve f(x) = 0.

2 (i) Show that $(x + 1)$ is a factor of the polynomial $h(x) = 2x^3 - 5x^2 - 4x + 3$.

(ii) Hence factorise h(x) fully.

(iii) Solve h(x) = 0.

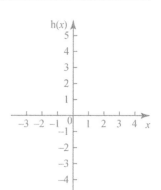

(iv) Sketch the graph of y = h(x).

3 Find the value of the constant k given that $(x + 5)$ is a factor of $x^3 + 4x^2 - 11x + k$.

4 When the polynomial f(x) = $x^4 - 6x^3 + 7x^2 + px + q$ is divided by $(x + 1)$ the remainder is zero.
   When f(x) is divided by $(x + 2)$ the remainder is 72. Find the values of p and q.

5 When the polynomials h(x) = $x^4 - 8x^3 + kx^2 - 6x - 9$ and g(x) = $kx^3 + 3x^2 + 7x + 13$ are divided by $(x + 1)$
   the remainders are the same. Find the value of k.

6 Let p(x) = $x^3 + ax^2 + 3x$ be a polynomial, where a is a real number.
   When p(x) is divided by $(x - 2)$ the remainder is 26. Find the remainder when p(x) is divided by $(x + 4)$.

1.3 The modulus function

1 (i) Sketch the graphs of the following on the grid provided.

$y = |x + 2|$

$y = |x| + 2$

$y = |2x|$

$y = 2|x|$

$y = -2|x|$

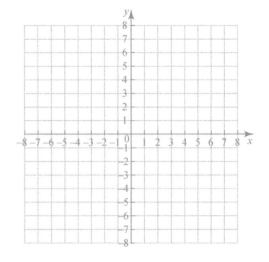

(ii) Describe what effect the constants a, b, c, d, and e have in these equations.

Equation	Description
$y = |x + a|$	
$y = |x| + b$	
$y = |cx|$	
$y = d|x|$	
$y = -e|x|$	

2 Sketch the following graphs.

(i) $y = |2x - 1|$

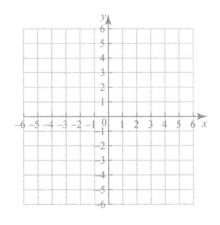

(ii) $y = |3 - x|$

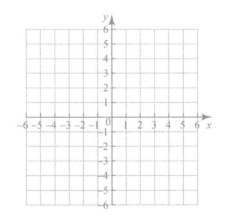

(iii) $y = 2|x + 1|$

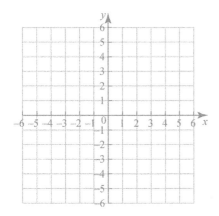

(iv) $y = |x^2 - 1|$

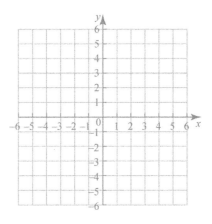

3 The graph of $y = \sin x$ is shown on the axes below for $-2\pi \leqslant x \leqslant 2\pi$.

On the axes provided, sketch the graphs of the equations given.

(i) $y = |\sin x|$

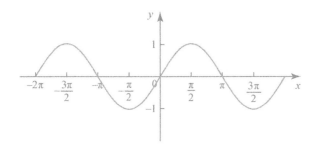

(ii) $y = \sin|x|$

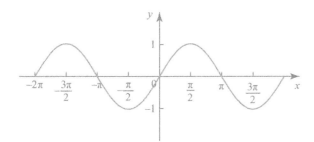

(iii) $y = \big|\sin|x|\big|$

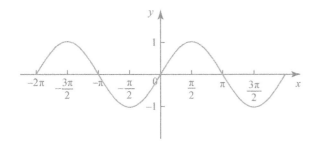

4 Solve the following equations.

(i) $|x - 1| = 2$

(ii) $|2x + 3| = 7$

(iii) $|x + 4| = |x - 2|$

(iv) $|3x + 4| = |4 - x|$

5 Solve the following inequalities.

(i) $|x + 2| < 5$

(ii) $|x + 1| \leqslant |x - 2|$

(iii) $|2x - 3| > |x + 3|$

(iv) $3|x + 2| < 1 - x$

6 (i) Solve the inequality $|2x + 1| \leqslant |x - 3|$.

(ii) Given that x satisfies the inequality $|2x + 1| \leqslant |x - 3|$, find the greatest possible value of $|x + 2|$.

OCR Core Mathematics 3 4723 Paper 01 Q5 June 2010

7   Given that a is a positive constant, solve the inequality $|x - 2a| < |x + a|$.

Further practice

1 Find the values of the constants A, B, C and D in the identity $2x^3 - 3x^2 + x - 2 \equiv (x + 2)(Ax^2 + Bx + C) + D$.

OCR MEI Further Concepts for Advanced Mathematics FP1 4755 Q2 June 2006

2 Simplify the following expressions as much as possible.

(i) $(3x^4 - x^3 + 6x - 11) - (2x^4 + x^2 - 4x - 12)$ (ii) $(x^2 - 1)^2 - (x^2 - 2)^2$

3 Use long division to divide the following polynomials, giving the quotient and remainder.

(i) $x - 1 \overline{\smash{)}\,x^3 - 4x + 3}$ (ii) $\dfrac{4x^3 - 7x^2 + 8x + 14}{x^2 - 2x + 1}$

4 Find the value of the constant p given that $(x - 4)$ is a factor of $x^3 - 3x^2 + px$.

5 Find the value of the constant m given that $(2x - 1)$ is a factor of $f(x) = 2x^4 - 5x^3 + mx^2 - x + 4$. Hence factorise $f(x)$ completely.

6 Find the values of p and q if $(x - 2)$ and $(x + 1)$ are factors of $f(x) = x^4 + px^3 + 9x^2 + qx - 12$.

7 The polynomial $2x^3 + 6x^2 + ax + b$, where a and b are constants, is denoted by $p(x)$. It is given that when $p(x)$ is divided by $(x + 3)$ the remainder is -25 and that when $p(x)$ is divided by $(x - 2)$ the remainder is 55.

(i) Find the values of a and b.

(ii) When a and b have these values, find the quotient and remainder when $p(x)$ is divided by $(x^2 + 1)$.

8 When $x^4 - 2x^3 - 7x^2 + 7x + a$ is divided by $(x^2 + 2x - 1)$, the quotient is $x^2 + bx + 2$ and the remainder is $cx + 7$. Find the values of the constants a, b and c.

9 Solve the following equations.

(i) $7 - |3 + 4x| = 2$ (ii) $|2x - 1| = |x + 3|$

10 Solve the following inequalities.

(i) $|3x - 1| \geqslant 8$ (ii) $|4x| \leqslant |3 - x|$

11 Solve the equation $|3x + 4a| = 5a$, where a is a positive constant.

OCR Core Mathematics 3 4723 Paper 01 Q1 January 2011

12 Solve $|a - x| < 2|x + 3a|$.

Past exam questions

1 Solve the inequality $2|x - 3| > |3x + 1|$.   [4]

Cambridge International AS & A Level Mathematics 9709 Paper 31 Q1 November 2010

2 The polynomial $x^3 + 4x^2 + ax + 2$, where a is a constant, is denoted by $p(x)$. It is given that the remainder when $p(x)$ is divided by $(x + 1)$ is equal to the remainder when $p(x)$ is divided by $(x - 2)$.

(i) Find the value of a.   [3]

(ii) When a has this value, show that $(x - 1)$ is a factor of $p(x)$ and find the quotient when $p(x)$ is divided by $(x - 1)$.   [3]

Cambridge International AS & A Level Mathematics 9709 Paper 23 Q3 November 2010

3 The polynomial $ax^3 - 3x^2 - 11x + b$, where a and b are constants, is denoted by $p(x)$. It is given that $(x + 2)$ is a factor of $p(x)$, and that when $p(x)$ is divided by $(x + 1)$ the remainder is 12.

(i) Find the values of a and b.   [5]

(ii) When a and b have these values, factorise $p(x)$ completely.   [3]

Cambridge International AS & A Level Mathematics 9709 Paper 22 Q7 November 2011

4 (i) Given that $(x + 2)$ is a factor of $4x^3 + ax^2 - (a + 1)x - 18$, find the value of the constant a. [3]

   (ii) When a has this value, factorise $4x^3 + ax^2 - (a + 1)x - 18$ completely. [3]

Cambridge International AS & A Level Mathematics 9709 Paper 23 Q2 June 2015

5 (i) Find the quotient and remainder when $2x^3 - 7x^2 - 9x + 3$ is divided by $x^2 - 2x + 5$. [3]

   (ii) Hence find the values of the constants p and q such that $x^2 - 2x + 5$ is a factor of $2x^3 - 7x^2 + px + q$. [2]

Cambridge International AS & A Level Mathematics 9709 Paper 22 Q2 June 2016

6 Solve the equation $|x + a| = |2x - 5a|$, giving x in terms of the positive constant a. [3]

Cambridge International AS & A Level Mathematics 9709 Paper 22 Q1 June 2017

▶ STRETCH AND CHALLENGE

· ·

1 When the polynomial $f(x) = x^3 + ax^2 + 12x + b$ is divided by $g(x) = x + 5$ the quotient is $x^2 + 10x + c$ and the remainder is 150.
   Find the values of a, b and c.

2 The solution to the inequality $|ax - b| \le |cx - d|$ is $0 \le x \le 2$, where a, b, c and d are constants.
   Find three different sets of values for a, b, c and d that make the inequality true.

3 Solve $\left|\dfrac{2x + 1}{x - 1}\right| \le 2$.

4 The formula for the roots of a general quadratic equation $ax^2 + bx + c = 0$ is well known as

$$x = \frac{-b \pm \sqrt{b^2 - 4ac}}{2a}.$$

A similar formula for the roots of a general cubic equation $ax^3 + bx^2 + cx + d = 0$ is more elusive!

(i) Given that we know one root of the cubic equation is $x = r$, show that

$$\frac{ax^3 + bx^2 + cx + d}{x - r} = ax^2 + (b + ar)x + (c + br + ar^2).$$

(ii) Hence solve the quadratic factor of the polynomial to obtain a formula for the other two roots of the cubic equation in terms of a, b, c and r.

5 The numbers a, b and c satisfy the equations

$a + b + c = 5$
$a^2 + b^2 + c^2 = 9$
$\dfrac{1}{a} + \dfrac{1}{b} + \dfrac{1}{c} = 2$

(i) Find the values of $ab + ac + bc$ and abc.

(ii) Show that a, b and c are the roots of the equation $x^3 - 5x^2 + 8x - 4 = 0$.

(iii) Find the values of a, b and c.

6 A polynomial $f(x)$ of degree 45 has a remainder of 4 when divided by $(x - 1)$, a remainder of 7 when divided by $(x - 2)$ and a remainder of 25 when divided by $(x - 4)$.
   Find the remainder when $f(x)$ is divided by $(x - 1)(x - 2)(x - 4)$.

2 Logarithms and exponentials

2.1 Exponential functions and logarithms

1 Write an equivalent statement involving logarithms for each of these.

(i) $2^4 = 16$

(ii) $3^3 = 27$

(iii) $4^{-2} = \dfrac{1}{16}$

2 Write an equivalent statement involving indices for each of these.

(i) $\log_2 \dfrac{1}{2} = -1$

(ii) $\log_3 9 = 2$

(iii) $\log_4 2 = \dfrac{1}{2}$

3 Find the value of each of the following (without using a calculator).

(i) $\log_2 8$

(ii) $\log_6 \dfrac{1}{216}$

(iii) $\log_{\frac{1}{3}} 27$

(iv) $\log_{16} 4$

(v) $\log_5 5\sqrt{5}$

(vi) $\log_{30} 30$

(vii) $\log_{\sqrt{2}} 4$

(viii) $\log_8 \sqrt{2}$

(ix) $\dfrac{\log 9}{\log 27}$

4 Find the value of the unknown in each of these equations.

(i) $\log_2 a = 5$

(ii) $\log_c 27 = 3$

(iii) $\log_5 \frac{1}{25} = e$

5 Evaluate these logarithms.

(i) $\log_a a$

(ii) $\log_c 1$

(iii) $\log_f \frac{1}{\sqrt{f}}$

6 Write each expression as the logarithm of a single number.

(i) $\log 5 + \log 4$

(ii) $\log 14 - \log 2$

(iii) $\frac{1}{2}\log 36$

(iv) $2\log 3 + 3\log 2$

(v) $\frac{1}{2}\log 100 - 2\log 5$

(vi) $\log 8 - \log 2 + \log 5$

7 Given that $a = \log 2$ and $b = \log 3$, write the following in terms of a and b.

(i) $\log 6$

(ii) $\log \frac{9}{4}$

(iii) $\log 0.\dot{2}$

8 Write the following equations without logarithms.

(i) $\log_{10} A = 2\log_{10} b + 1$

(ii) $2\log_5 D = \log_5(E - 1) + \log_5 3$

9 Solve these equations.

(i) $2^x = 12$

(ii) $6^{2x} = 4^{x-1}$

(iii) $2^{x-1}3^x = 16$

(iv) $\log_{10} x + 3 = \log_{10}(x + 3)$

(v) $\log_2(x - 1) = \log_2 x - 1$

(vi) $7^{x+2} = 7^x + 7^2$

10 Solve the following inequalities.

(i) $2^x < 5$

(ii) $3 \times 2^{3x-1} < 5$

11 Find the values of a and b from the graphs below.

(i) $y = a(2^x) + b$

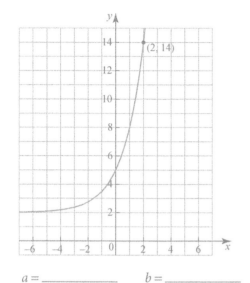

(ii) $y = \log_{10}(x + a) + b$

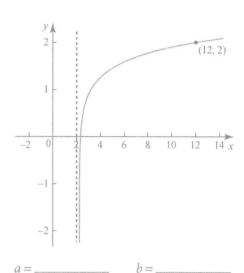

$a = $ _____ $b = $ _____ $a = $ _____ $b = $ _____

12 The magnitude (M) of an earthquake is measured on the Richter scale using the formula $M = \log_{10} \dfrac{I}{S}$, where I is the intensity of the earthquake and S is the intensity of a 'standard' earthquake.

In 2010 an earthquake in Christchurch, New Zealand, registered 7.1 on the Richter scale and in 1985 Mexico City experienced an 8.3 magnitude earthquake.

(i) How many times greater was the intensity of the Mexico City earthquake than the New Zealand earthquake? Give your answer to the nearest whole number.

(ii) Find the magnitude of an earthquake that would be half the intensity of the earthquake in Mexico City.

(iii) Find the magnitude of an earthquake that would be double the intensity of the earthquake in Christchurch.

13 Solve the following equations simultaneously.

$$2\log_{10} x + \log_{10} y = 2$$
$$xy^2 = 80$$

2.2 Modelling curves

1 **(i)** Show that the equation $y = kx^p$ can be written in the form

$$\log y = p\log x + \log k$$

(ii) Hence state the gradient and y-intercept of the straight line on the graph of $\log y$ against $\log x$.

2 The average weight loss (W kg) of a large group of people on a diet is measured after 1, 2, 5 and 10 months (m) on the diet.

It is proposed that the average weight loss after m months on the diet can be modelled by an equation of the form $W = Am^b$, where A and b are constants.

(i) Complete the table.

m	W	$\log_{10} m$	$\log_{10} W$
1	8.00		
2	5.66		
5	3.58		
10	2.53		

(ii) The graph of $\log_{10} m$ against $\log_{10} W$ is shown on the right.

Use the graph or the table to determine the values of the constants A and b.

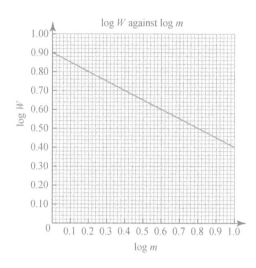

(iii) Based on this model:

(a) calculate the average weight loss at 12 months

(b) find when the average weight loss will be less than 1 kg.

3 The average length (L cm) of male babies born in Karachi is measured at regular intervals. Let t be the age in months of the babies.

A doctor proposed that the data can be modelled with an equation of the form $L = Ab^t$, where A and b are constants.

(i) Complete the table.

(ii) Show that the equation $L = Ab^t$ can be reduced to linear form by taking logarithms of both sides.

t	L	$\log_{10} L$
0	50	
5	57	
10	66	
15	77	
20	87	
36	128	

(iii) The graph of $\log_{10} L$ against t is shown here.
Use the graph or the table to determine the values of A and b.

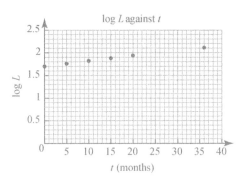

(iv) What does the model predict the height of an average 18-year-old man will be?
Is your answer reasonable? Why or why not?

4 The variables x and y satisfy the equation $y = kb^{-x}$, where k and b are constants.

The graph of $\log_{10} y$ against x is a straight line passing through the points (1.3, 3.4) and (4.2, 0.7).

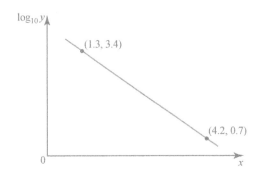

Find the values of k and b correct to 3 significant figures.

2.3 The natural logarithm and exponential functions

1 Solve the following equations, giving your answers exactly.

(i) $e^{2x+1} = 4$

(ii) $\ln x - \ln 4 = \ln(x - 4)$

(iii) $e^{2x} + e^{x} = 30$

(iv) $\ln x = \ln(x + 1) + 1$

2 Write the equation $\ln A = 3\ln B + 2\ln 3$ without logarithms.

3 Simplify the following expressions.

(i) $\ln(e^{x+y})^2$

(ii) $e^{2\ln x + 3\ln y}$

4   The variables x and y satisfy the equation $y = Ax^b$.
    Given that the graph of $\ln y$ against $\ln x$ is a straight line passing
    through $(1.2, 0.5)$ and $(8.1, 12.3)$, find the values of the constants A and b.

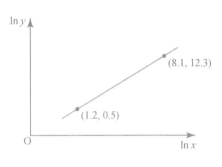

5   The number of bacteria in a colony, N, can be modelled by the
    equation $N = 1000e^{0.4t}$, where t is the time in hours since
    measurements were started.

    (i)   What is the initial size of the colony?

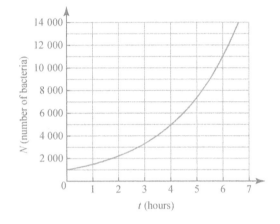

    (ii) Find the number of bacteria after 5 hours.

    (iii) Calculate how long (to the nearest minute) it took for the bacteria to double in number.

    (iv) Find how many hours it would take for the number in the colony to pass 1 million bacteria.

6 Find the x-coordinate of the point of intersection of the two curves $y = e^{x-1}$ and $y = e^{-x}$.

7 It is given that $p = e^{280}$ and $q = e^{300}$.

   (i) Use logarithm properties to show that $\ln\left(\dfrac{ep^2}{q}\right) = 261$.

   (ii) Find the smallest integer n which satisfies the inequality $5^n > pq$.

OCR Core Mathematics 3 4723 Paper 01 Q2 June 2012

Further practice

1 Find the value of each of the following (without using a calculator).

   (i) $\log_{10} 100$

   (ii) $\log_{10} 0.1$

   (iii) $\log_7 1$

   (iv) $\dfrac{\log 8}{\log 2}$

   (v) $\log_9 \sqrt{3}$

   (vi) $\dfrac{\log 0.2}{\log 25}$

2 Find the value of the unknown in each of the following equations.

   (i) $\log_3 b = -2$

   (ii) $\log_d 8 = \dfrac{1}{2}$

   (iii) $\log_{36} 6 = f$

3 Evaluate the following logarithms.

   (i) $\log_b b^2$

   (ii) $\log_d \sqrt[3]{d}$

   (iii) $\log_e \dfrac{1}{e}$

4 Write each of the following expressions as the logarithm of a single number.

   (i) $\log 12 + \log 12$

   (ii) $3\log 4$

   (iii) $2\log 4 - \log 32$

   (iv) $\dfrac{1}{2}\log 9 - \dfrac{1}{3}\log 8 + \dfrac{1}{4}\log 256$

   (v) $2\log 5 + \dfrac{1}{3}\log 64 - \dfrac{1}{2}\log 121$

5 Given that $a = \log 2$ and $b = \log 3$, write the following in terms of a and b.

   (i) $\log 12$

   (ii) $\log 8$

   (iii) $\log \dfrac{4}{\sqrt{3}}$

6 Write the following equations without logarithms.

(i) $\log_{10} M = \log_{10}\left(\dfrac{P}{Q}\right) - \log_{10} R^2$ (ii) $\log_3 S = 2 - \log_3 \sqrt{T} - 3\log_3 U$

7 Solve the following equations.

(i) $3^{x+1} = 24$ (ii) $3^{x+1} = 4^{2x-1}$ (iii) $\log_5(x-1) = \log_5 x - 1$

(iv) $\log_3(x+4) - \log_3(x-4) = 1$ (v) $3 - 4^x = \dfrac{2}{4^x}$

8 Solve the following inequalities.

(i) $4^{x+2} - 2 < 18$ (ii) $\left|3^x - 2\right| \geqslant 1$

9 The loudness of a sound (L) is measured in decibels (dB) according to the formula $L = 20\log_{10}\left(\dfrac{P}{P_0}\right)$, where P is the power (or intensity) of the sound and P_0 is a fixed reference power.
A rock band registers at 110 dB and a plane taking off is 125 dB.
How many times greater is the intensity of the sound of the plane compared to the rock band?

10 Solve the following equations, giving your answers exactly.

(i) $\ln 2x = \dfrac{1}{2}\ln 16 + \dfrac{2}{3}\ln 8$ (ii) $3e^{-x} + 1 = 2e^x$

(iii) $\ln(x+1) + 1 = 3$ (iv) $e^{3x} - e^{2x} = 2e^x$

11 (i) Show that the equation $\log_4(x-4) = 2 - \log_4 x$ can be written as a quadratic equation in x.

(ii) Hence solve the equation $\log_4(x-4) = 2 - \log_4 x$, giving your answer to 2 decimal places.

12 Solve the equation $\left|8 - 3^x\right| = 15$.

13 The variables x and y are related by the equation $y = Ab^x$ where A and b are constants. The graph of $\ln y$ against x is a straight line passing through the points (4.1, 4.5) and (5.9, 9.4), as shown.
Find the values of A and b to 3 significant figures.

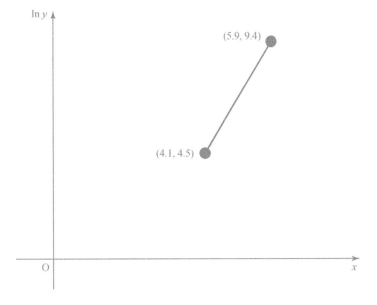

14 Solve the equation $6^x = 6^{x-1} + 6$, giving your answer correct to 3 significant figures.

15 Solve the equation $\ln(3x+2) = 2\ln x + \ln 2$.

16 Solve the equation $5^{x-1} = 7^{2x-1}$.

17 The equation of the line of best fit for a set of data on the graph of x against $\log_{10} y$ is $\log_{10} y = 0.7x + 3.5$.
Find a suitable model for the data in the form $y = A(b)^x$.

Past exam questions

1 (i) Given that $y = 2^x$, show that the equation $2^x + 3(2^{-x}) = 4$ can be written in the form $y^2 - 4y + 3 = 0$. [3]

   (ii) Hence solve the equation

   $$2^x + 3(2^{-x}) = 4,$$

   giving the values of x correct to 3 significant figures where appropriate. [3]

   Cambridge International AS & A Level Mathematics 9709 Paper 21 Q5 June 2010

2 Solve the equation $\ln(1 + x^2) = 1 + 2\ln x$,

   giving your answer correct to 3 significant figures. [4]

   Cambridge International AS & A Level Mathematics 9709 Paper 31 Q2 November 2010

3 The variables x and y satisfy the equation $y = A(b^x)$, where A and b are constants. The graph of $\ln y$ against x is a straight line passing through the points $(1.4, 0.8)$ and $(2.2, 1.2)$, as shown in the diagram. Find the values of A and b, correct to 2 decimal places. [6]

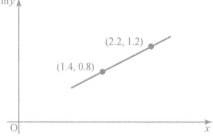

   Cambridge International AS & A Level Mathematics 9709 Paper 23 Q5 November 2010

4 Use logarithms to solve the equation $3^x = 2^{x+2}$, giving your answer correct to 3 significant figures. [4]

   Cambridge International AS & A Level Mathematics 9709 Paper 22 Q1 June 2011

5 (a) Find the value of x satisfying the equation $2\ln(x - 4) - \ln x = \ln 2$. [5]

   (b) Use logarithms to find the smallest integer satisfying the inequality $1.4^y > 10^{10}$. [3]

   Cambridge International AS & A Level Mathematics 9709 Paper 22 Q4 November 2014

6 Using the substitution $u = 3^x$, solve the equation $3^x + 3^{2x} = 3^{3x}$ giving your answer correct to 3 significant figures. [5]

   Cambridge International AS & A Level Mathematics 9709 Paper 31 Q2 November 2015

7 Given that $5^{3x} = 7^{4y}$, use logarithms to find the value of $\frac{x}{y}$ correct to 4 significant figures. [3]

   Cambridge International AS & A Level Mathematics 9709 Paper 22 Q1 June 2016

8 The variables x and y satisfy the equation $y = \dfrac{K}{a^{2x}}$, where K and a are constants.

   The graph of $\ln y$ against x is a straight line passing through the points $(0.6, 1.81)$ and $(1.4, 1.39)$, as shown in the diagram.

   Find the values of K and a correct to 2 significant figures. [6]

   Cambridge International AS & A Level Mathematics 9709 Paper 22 Q5 June 2017

STRETCH AND CHALLENGE

1 Evaluate these expressions.

(i) $\log_4 \sqrt[5]{\dfrac{32}{1024}}$

(ii) $\log_2\left[\dfrac{\sqrt{256}\left(\frac{1}{8}\right)^6}{32\left(\frac{1}{2}\right)^3}\right]$

(iii) $e^{\ln e^{\ln \pi}}$

(iv) $36^{\frac{1}{2}-\log_6 \sqrt{3}}$

2 For the following equations state whether they are

- always true
- sometimes true
- never true.

Explain your reasoning.

(i) $\log_a b = \log_b a$

(ii) $\log_{\frac{1}{a}} a = 1$

(iii) $\log_a a = 0$

(iv) $\log_a(\log_a a) = 1$

(v) $\dfrac{\log a}{\log b} = \log a - \log b$

(vi) $\log_a(\log_b(\log_c c)) = 0$

(vii) $\log_a x + \log_b x = \log_{ab} x$

3 Solve the equation $3(2^x) - 4^x = 2$.

4 (i) Prove that $\log_a b = \dfrac{\log_c b}{\log_c a}$ (this is called the *change of base law*).

(ii) Hence show that $\log_a b = \dfrac{1}{\log_b a}$.

5 Solve $\log_3 x + \log_4 x - 1 = \log_5 x$.

6 Show that if $3\log_x y + 3\log_y x = 10$ then $y = x^3$ or $x = y^3$.

7 Given that $2\log(x - 2y) = \log x + \log y$, find the possible values of $\dfrac{x}{y}$.

8 The variables x and y are related so that when xy is plotted against x^2, the result is a straight line passing through the points $(4, 6)$ and $(9, 21)$ as shown.

(i) Find the value of y when $x = 6$.

(ii) Find the two possible values of x when $y = 7$.

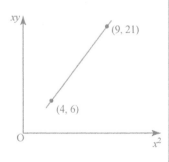

9 Solve the equation $3^{\log_{10} x^2} = 2(3^{1+\log_{10} x}) + 27$.

3 Trigonometry

3.1 Reciprocal trigonometrical functions

1 The graphs of $y = \sin x$, $y = \cos x$ and $y = \tan x$ for $-2\pi \leqslant x \leqslant 2\pi$ are shown below. Use these to sketch the following graphs.

(i) $y = \operatorname{cosec} x = \dfrac{1}{\sin x}$
    (ii) $y = \sec x = \dfrac{1}{\cos x}$

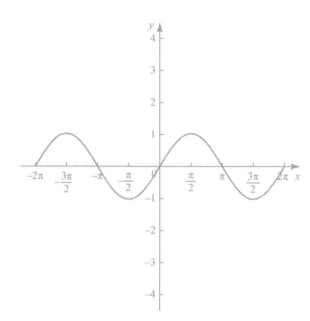
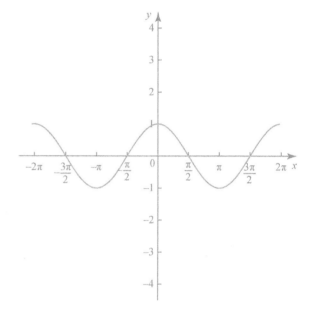

(iii) $y = \cot x = \dfrac{1}{\tan x} = \dfrac{\cos x}{\sin x}$

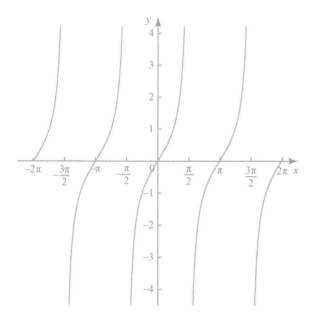

2 Without using a calculator, find the exact value of the following.

(i) $\operatorname{cosec} 150°$

(ii) $\sec \dfrac{\pi}{4}$

(iii) $\cot 300°$

(iv) $\operatorname{cosec} \dfrac{4\pi}{3}$

3 Eliminate θ from these equations.

(i) $x = 2\operatorname{cosec}\theta \qquad y = 3\cot\theta$

(ii) $x = \sin\theta - \cos\theta \qquad y = \sin\theta + \cos\theta$

4 Solve the following equations over the given domains.

(i) $\operatorname{cosec}\theta = 4$ for $0° \leqslant \theta \leqslant 360°$

(ii) $\tan\theta + \cot\theta = -4$ for $-\pi \leqslant \theta \leqslant \pi$

5 Given that $\sec\theta = 3$ and $0° \leqslant \theta \leqslant 90°$, find the **exact** values of the following.

(i) $\cos\theta$

(ii) $\sin\theta$

(iii) $\operatorname{cosec}\theta$

(iv) $\cot\theta$

6   Prove these identities.

(i) $\dfrac{1}{\tan\theta + \cot\theta} \equiv \sin\theta\cos\theta$

(ii) $\sec^4\theta - \tan^4\theta \equiv \sec^2\theta + \tan^2\theta$

7   $m = \dfrac{1 + \cos\theta}{\sin\theta}$

(i)   Show that $\dfrac{1}{m} = \dfrac{1 - \cos\theta}{\sin\theta}$.

(ii)   Find an expression for $\cos\theta$ in terms of m only.

3.2 Compound-angle formulae

1   Find the exact value of each of the following.

(i) $\cos 75° = \cos(45° + 30°)$

(ii) $\sin 15° = \sin(60° - 45°)$

2   Simplify each of the following expressions as much as possible.

   (i) $\sin(\theta - 30°)$

   (ii) $\cos\left(\dfrac{\pi}{4} - \theta\right)$

3   Write each of the following expressions in the form $\sin(A \pm B)$ or $\cos(A \pm B)$.

   (i) $\sin\theta\cos2\beta + \sin2\beta\cos\theta$

   (ii) $\cos3\theta\cos\theta + \sin3\theta\sin\theta$

4   Use the diagram to find the exact value of $\sin(x + y)$.

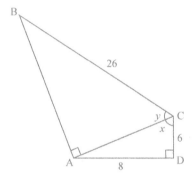

5   The angles P and Q are both acute with $\cos P = \dfrac{2}{5}$ and $\tan Q = \dfrac{7}{3}$.

   Find the **exact** value of each of the following.

   (i) $\cos(P - Q)$

   (ii) $\sin(P + Q)$

6   Solve these equations.

   (i) $\sin(45° - \theta) = \cos(30° + \theta)$ for $-180° \leqslant \theta \leqslant 180°$

   (ii) $\tan\left(\dfrac{\pi}{3} + \theta\right) = 2\tan\left(\dfrac{\pi}{6} - \theta\right)$ for $0 \leqslant \theta \leqslant \pi$

7   Prove the identity $\cos(A + B)\cos(A - B) \equiv \cos^2 A - \sin^2 B$.

8   *A* is an acute angle and *B* is an obtuse angle such that $\tan A = \frac{1}{3}$ and $\tan(A - B) = 5$.

(i) Find $\tan B$.

(ii) Hence show that the exact value of $\sin(A + B)$ is $\dfrac{17}{\sqrt{650}}$.

3.3 Double-angle formulae

1   Use the double-angle formulae to find the exact value of these.

(i) $\sin\dfrac{2\pi}{3}$

(ii) $\cos\dfrac{2\pi}{3}$

(iii) $\tan\dfrac{2\pi}{3}$

2 Given θ from the triangle shown, find the exact value of these.

(i) $\sin 2\theta$ (ii) $\cos 2\theta$

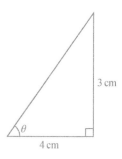

3 cm

4 cm

θ

3 Given that $\sin\theta = -\dfrac{2}{3}$ and $\dfrac{3\pi}{2} \leqslant \theta \leqslant 2\pi$, find the exact value of the following.

(i) $\sin 2\theta$ (ii) $\cos 2\theta$ (iii) $\sin 4\theta$

4 (i) Given that $5 + 4\sec^2\theta = 12\tan\theta$, find the exact value of $\tan\theta$.

(ii) Hence find the exact value of these expressions.

(a) $\tan(\theta + 45°)$ (b) $\tan 2\theta$

5   Solve these equations.

(i) $\sin 2\theta = \sin \theta$ for $0 \le \theta \le 2\pi$

(ii) $3\tan 2\theta + 2\tan \theta = 0$ for $-180° < \theta < 180°$

(iii) $\dfrac{1 - \sin\theta - \cos 2\theta}{\cos\theta - \sin 2\theta} = 1$ for $0° \le \theta \le 360°$

6   Prove the following identities.

(i) $\sin(45° + \theta)\sin(45° - \theta) \equiv \dfrac{1}{2}\cos 2\theta$

(ii) $\dfrac{2\sin\theta\cos\theta}{\cos^4\theta - \sin^4\theta} \equiv \tan 2\theta$

7 (i) Prove the identity $\cot\theta + \tan\theta \equiv 2\operatorname{cosec} 2\theta$.

(ii) Hence solve the equation $\cot\theta + \tan\theta = 8$ for $0 \leqslant \theta \leqslant 2\pi$.

8 Given that $\sin 25° = k$, where k is a positive constant, express the following in terms of k.

(i) $\sin 50°$ (ii) $\cos 50°$ (iii) $\tan 155°$

3.4 The forms $r\cos(\theta \pm \alpha)$, $r\sin(\theta \pm \alpha)$

1   Write these expressions in the form given, where $r > 0$ and $0° < \alpha < 90°$.

(i) $\sin\theta - 3\cos\theta$ in the form $r\sin(\theta - \alpha)$

(ii) $12\cos\theta + 5\sin\theta$ in the form $r\cos(\theta - \alpha)$

2   (i)   Express $2\cos\theta + \sin\theta$ in the form $r\cos(\theta - \alpha)$, where $r > 0$ and $0° < \alpha < 90°$.

(ii) Hence solve $2\cos\theta + \sin\theta = 1$ for $0° \leqslant \theta \leqslant 360°$.

(iii) Sketch the graph of $y = 2\cos\theta + \sin\theta$ on the axes provided.

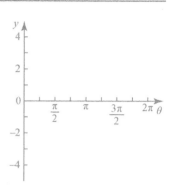

(iv) Find the greatest and least values of $2\cos\theta + \sin\theta + 5$ as θ varies.

3 The function $T(\theta)$ is defined for θ in degrees by $T(\theta) = 3\cos(\theta - 60°) + 2\cos(\theta + 60°)$.

(i) Express $T(\theta)$ in the form $A\sin\theta + B\cos\theta$, giving the exact values of the constants A and B.

(ii) Hence express $T(\theta)$ in the form $R\sin(\theta + \alpha)$, where $R > 0$ and $0° < \alpha < 90°$.

(iii) Find the smallest positive value of θ such that $T(\theta) + 1 = 0$.

4 (i) Express $3\sin\theta + 4\cos\theta$ in the form $R\sin(\theta + \alpha)$, where $R > 0$ and $0° < \alpha < 90°$.

(ii) Hence

(a) solve the equation $3\sin\theta + 4\cos\theta + 1 = 0$, giving all solutions in the interval $-180° < \theta < 180°$.

(b) find the values of the positive constants k and c such that

$$-37 \leqslant k(3\sin\theta + 4\cos\theta) + c \leqslant 43$$

for all values of θ.

OCR Core Mathematics 3 4723 Paper 01 Q8 June 2012

Further practice

1   Solve the equation $\sin(x - 45°) - \cos(45° - x) = 1$ for $0° \leqslant x \leqslant 360°$.

2   Solve the equation $2\sin(x + 30°) = \cos(x - 45°)$, giving all solutions in the interval $0° \leqslant \theta \leqslant 180°$.

3   (i)   Prove the identity $2\operatorname{cosec} 2\theta \equiv \sec\theta \operatorname{cosec}\theta$.

   (ii)   Hence solve the equation $\sec\theta \operatorname{cosec}\theta = 4$ for $0° \leqslant \theta \leqslant 180°$.

4   Solve the equation $4\operatorname{cosec}^2\theta - 7 = 4\cot\theta$ for $0° \leqslant \theta \leqslant 180°$.

5   (i)   Express $8\sin\theta + 15\cos\theta$ in the form $R\sin(\theta + \alpha)$, where $R > 0$ and $0° < \alpha < 90°$.

   (ii)   Hence solve $8\sin\theta + 15\cos\theta = 14$ for $0° \leqslant \theta \leqslant 360°$.

   (iii)   Find the range of values of the constant k such that the equation $8\sin\theta + 15\cos\theta = k$ has no solutions.

6   (i)   Prove the identity $\tan(\theta + 60°)\tan(\theta - 60°) = \dfrac{\tan^2\theta - 3}{1 - 3\tan^2\theta}$.

   (ii)   Solve, for $0° < \theta < 180°$, the equation $\tan(\theta + 60°)\tan(\theta - 60°) = 4\sec^2\theta - 3$, giving your answers correct to the nearest $0.1°$.

   (iii)   Show that, for all values of the constant k, the equation $\tan(\theta + 60°)\tan(\theta - 60°) = k^2$ has two roots in the interval $0° < \theta < 180°$.

OCR Core Mathematics 3 4723 Paper 01 Q9 June 2007

7   (i)   Express $\tan 2\alpha$ in terms of $\tan\alpha$ and hence solve, for $0° < \alpha < 180°$, the equation $\tan 2\alpha \tan\alpha = 8$.

   (ii)   Given that β is the acute angle such that $\sin\beta = \dfrac{6}{7}$, find the exact value of:

     (a) $\operatorname{cosec}\beta$ (b) $\cot^2\beta$.

OCR Core Mathematics 3 4723 Paper 01 Q5 June 2008

Past exam questions

1   (i)   Show that the equation $\tan(x + 45°) = 6\tan x$ can be written in the form $6\tan^2 x - 5\tan x + 1 = 0$. [3]

   (ii)   Hence solve the equation $\tan(x + 45°) = 6\tan x$, for $0° < x < 180°$. [3]

Cambridge International AS & A Level Mathematics 9709 Paper 21 Q3 June 2010

2   Solve the equation $\cos(\theta + 60°) = 2\sin\theta$, giving all solutions in the interval $0° \leqslant \theta \leqslant 360°$. [5]

Cambridge International AS & A Level Mathematics 9709 Paper 31 Q3 November 2010

3   (i)   Prove that $\sin^2 2\theta(\operatorname{cosec}^2\theta - \sec^2\theta) = 4\cos 2\theta$. [3]

   (ii)   Hence

     (a)   solve for $0° \leqslant \theta \leqslant 180°$ the equation $\sin^2 2\theta(\operatorname{cosec}^2\theta - \sec^2\theta) = 3$, [4]

     (b)   find the exact value of $\operatorname{cosec}^2 15° - \sec^2 15°$. [2]

Cambridge International AS & A Level Mathematics 9709 Paper 22 Q8 June 2011

4   Express the equation $\cot 2\theta = 1 + \tan\theta$ as a quadratic equation in $\tan\theta$. Hence solve this equation for $0° < \theta < 180°$. [6]

Cambridge International AS & A Level Mathematics 9709 Paper 33 Q3 November 2016

5 (i) Express $\sin 2\theta(3\sec\theta + 4\operatorname{cosec}\theta)$ in the form $a\sin\theta + b\cos\theta$, where a and b are integers. [3]

   (ii) Hence express $\sin 2\theta(3\sec\theta + 4\operatorname{cosec}\theta)$ in the form $R\sin(\theta + \alpha)$ where $R > 0$ and $0° < \alpha < 90°$. [3]

   (iii) Using the result of part (ii), solve the equation $\sin 2\theta(3\sec\theta + 4\operatorname{cosec}\theta) = 7$ for $0° \leqslant \theta \leqslant 360°$. [4]

   Cambridge International AS & A Level Mathematics 9709 Paper 23 Q7 November 2016

6 (i) Express the equation $\cot\theta - 2\tan\theta = \sin 2\theta$ in the form $a\cos^4\theta + b\cos^2\theta + c = 0$, where a, b, and c are constants to be determined. [3]

   (ii) Hence solve the equation $\cot\theta - 2\tan\theta = \sin 2\theta$ for $90° < \theta < 180°$. [2]

   Cambridge International AS & A Level Mathematics 9709 Paper 32 Q3 June 2017

▶ ## STRETCH AND CHALLENGE
• •

1 A projectile is fired from a sloping hill that makes an angle A with the horizontal. It is fired with velocity V m/s at an angle B to the hill as shown.

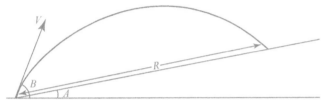

The range, R, that the projectile can travel is given by

$$R = \frac{2V^2 \sin B}{g\cos^2 A}\cos(A + B)$$

where g = acceleration due to gravity = 10 m/s^2.

   (i) Express R as a function of B given that $A = \dfrac{\pi}{4}$.

   (ii) If the projectile is fired at a speed of 40 m/s, find the angle B it should be fired at to hit a target with a range of 150 m.

2 For all values of x for which the terms are defined, it is given that

$$\tan x - \tan\frac{1}{8}x = \frac{\sin kx}{\cos x \cos\frac{1}{8}x}.$$

   Find the value of the constant k.

3 Given that $\sin A + \cos A = 1.5$, find the value of $\sin^3 A + \cos^3 A$.

4 Given that $\cos\theta = 0.1$ and $0 \leqslant \theta \leqslant \dfrac{\pi}{2}$, find the value of $\log_{10}(\tan\theta) - \log_{10}(\sin\theta)$.

5 Prove the identity $1 - \dfrac{\sin^2\theta}{1 + \cot\theta} - \dfrac{\cos^2\theta}{1 + \tan\theta} \equiv \dfrac{1}{2}\sin 2\theta$.

6 An amusement park has a giant double Ferris wheel as shown below.

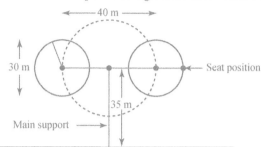

The double Ferris wheel has a rotating arm 40 metres long attached at its centre to a main support 35 metres above the ground. At each end of the rotating arm is attached a Ferris wheel measuring 30 metres in diameter, as shown in the diagram. The rotating arm takes 4 minutes to complete one full revolution, and each wheel takes 3 minutes to complete a revolution about that wheel's hub. All revolutions are anticlockwise, in a vertical plane.

At time $t = 0$ the rotating arm is parallel to the ground and your seat is at the 3 o'clock position of the rightmost wheel.

Find a formula for h(t), your height above the ground in metres, as a function of time in minutes.

NZQA Scholarship Calculus Q2b (i) 2007

7 A rectangular piece of paper of width 8 cm has one corner folded down so the corner rests against the opposite longer side as shown.

Show that $x = \dfrac{4}{\sin\theta\cos^2\theta}$.

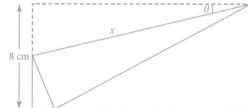

8 (i) Express $\cos 4\theta$ in terms of $\sin 2\theta$ and hence show that $\cos 4\theta$ can be expressed in the form

$1 - k\sin^2\theta\cos^2\theta$, where k is a constant to be determined.

(ii) Hence find the exact value of $\sin^2\left(\dfrac{1}{24}\pi\right)\cos^2\left(\dfrac{1}{24}\pi\right)$.

(iii) By expressing $2\cos^2 2\theta - \dfrac{8}{3}\sin^2\theta\cos^2\theta$ in terms of $\cos 4\theta$, find the greatest and least possible values of

$2\cos^2 2\theta - \dfrac{8}{3}\sin^2\theta\cos^2\theta$ as θ varies.

OCR Core Mathematics 3 4723 Paper 01 Q8 January 2012

4 Differentiation

4.1 The product rule

1 Differentiate the following functions.

(i) $y = (x - 2)(x + 3)^2$

(ii) $y = x^3(1 - 2x)^4$

(iii) $y = 3x^2\sqrt{1 + 4x}$

2 Find the equation of the tangent to the curve $f(x) = 4x(x - 3)^5$ at the point $(4, 16)$.

3 Find the x-values of the stationary points on the curve $h(x) = x^2(x + 3)^3$.

4 A curve is given by $f(x) = (x - 1)^k(x + 2)^{k+1}$, where k is a positive constant.

 (i) Two stationary points on the curve are at $x = -2$ and $x = 1$.
 Find the x-coordinate of the third stationary point in terms of k.

 (ii) The third stationary point occurs when $x = -\frac{1}{3}$. Calculate the value of k.

4.2 The quotient rule

1 Differentiate the following.

(i) $y = \dfrac{x^3}{x-1}$

(ii) $y = \dfrac{x+2}{3x^2}$

(iii) $y = \dfrac{2x}{\sqrt{6x-1}}$

2 Find the equation of the normal to the curve $g(x) = \dfrac{2x-1}{1-x^2}$ at the point $(2, -1)$.

3 The graph shows the curve $y = \dfrac{x^2 - x + 2}{x + 1}$.

   (i) Find $\dfrac{\mathrm{d}y}{\mathrm{d}x}$.

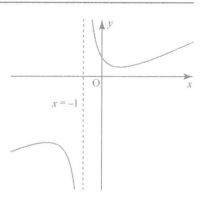

   (ii) Find the coordinates of the stationary points on the curve.

4.3 Differentiating natural logarithms and exponentials

1 Differentiate the following, simplifying your answers as much as possible.

(i) $y = \ln(x + 4)$

(ii) $y = \ln(3x^2)$

(iii) $y = e^{1-x}$

(iv) $y = 3e^{x^3+1}$

(v) $y = 3\ln\left(\dfrac{x}{x-1}\right)$

(vi) $y = 5\ln(1 + \sqrt{x})^2$

(vii) $y = xe^{x-2}$

(viii) $y = \dfrac{x^3}{e^x}$

(ix) $y = \dfrac{1 + e^x}{1 - e^{-x}}$

(x) $y = \sqrt{e^x}\,\ln x^2$

(xi) $y = \dfrac{e^{3x} - 1}{\ln(3x - 1)}$

(xii) $y = \ln(1 + e^{2x})e^{5x}$

2   The value of a car, $V, is given by the formula $V = 24\,000\,e^{-0.04t}$, where t is years since the car was bought.

   (i) At what rate is the price of the car changing after 6 years?

   (ii) Find the age of the car when the value is decreasing at a rate of $594/year.

3   The diagram shows the graph of $y = \dfrac{x}{\ln x}$ for $x > 1$.

   (i) Find $\dfrac{dy}{dx}$ and $\dfrac{d^2y}{dx^2}$.

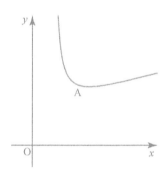

   (ii) Hence

      (a) find the exact coordinates of the stationary point A on the curve

      (b) find the exact coordinates of the point where the gradient is a maximum.

4 The graph of $y = x^2 e^{-x}$ is shown.

(i) Find $\dfrac{dy}{dx}$ and $\dfrac{d^2y}{dx^2}$.

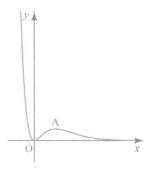

(ii) Find the exact coordinates of the stationary point A on the curve.

(iii) Find the x-values of the two points on the curve where $\dfrac{d^2y}{dx^2} = 0$. What do these points represent on the graph?

5 Find the exact coordinates of the stationary point on the curve $y = x^3 \ln x$ for $x > 0$.

4.4 Differentiating trigonometrical functions

1 Differentiate the following.

(i) $y = 3\sin 2x$

(ii) $y = \cos(1 + 3x)$

(iii) $y = \tan(x^2)$

(iv) $y = x^3 \sin 2x$

(v) $y = \dfrac{\cos 5x}{x^2}$

(vi) $y = \dfrac{\sin^3 x}{x^2}$

(vii) $y = e^{\sin x + 1}$

(viii) $y = \ln(\tan 2x)$

(ix) $y = \sin(\ln 3x)$

(x) $y = \tan\left(e^{x^2}\right)$

(xi) $y = \sqrt{\cos 2x}$

(xii) $y = \sin^3(2e^{x-1})$

(xiii) $y = \cos^4[\ln(\sin e^x)]$

(xiv) $y = e^{\sin^2(\ln x)}$

2 (i) Differentiate $y = x^2 \sin x$.

(ii) Hence find the equation of the tangent to the curve at $x = \pi$.

3 Given that $f(x) = 2\sin^2 3x$, find the exact value of $f'\left(\dfrac{\pi}{18}\right)$.

4 Consider the equation of a curve, $y = \dfrac{e^{2x}}{\cos x}$, for $x > 0$.

(i) Find $\dfrac{dy}{dx}$.

(ii) Hence find the x-coordinate of the stationary point of the curve for $0 \leqslant x \leqslant \pi$.

5 The diagram shows the curve $y = 3\sin^2 x \cos^3 x$ for $0 \leqslant x \leqslant \dfrac{\pi}{2}$ and its maximum point N.
Find the x-coordinate of N.

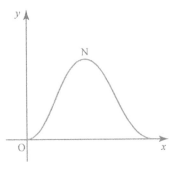

6 The curve $y = \dfrac{\cos 2x}{e^{2x}}$ has two stationary points for $0 \leqslant x \leqslant \pi$. Find the x-coordinates of these stationary points.

7 Consider the curve defined by $y = \ln(\cos 2x) + 2x$ for $0 \leqslant x \leqslant 2\pi$.

   (i) For what values of x is the function undefined for $0 \leqslant x \leqslant 2\pi$?

   (ii) Find the x-coordinates of all the stationary points on the curve.

   (iii) Determine the nature of each of these stationary points.

4.5 Differentiating functions defined implicitly

1 Differentiate the following with respect to x.

    (i) $2y^3$

    (ii) $3x^2 - 5y^4 - 8$

    (iii) $\sin 2x + \cos 2y$

    (iv) e^{3y}

    (v) $4x^2 y$

    (vi) $\ln(xy)$

    (vii) $\tan xy^2 - e^y$

    (viii) $(e^{\sin y})^x$

    (ix) $\dfrac{x \sin y}{y \sin x}$

    (x) $y^3 \ln(\sin xy)$

2 Differentiate with respect to x and find an expression for $\dfrac{dy}{dx}$ in terms of x and y.

(i) $y^2 - 2x^3 = 5$

(ii) $x^2 y = \sin y$

(iii) $2y^3 + x = 4xy$

(iv) $e^{xy} - 2x = 12$

3 Find the equation of the tangent to the curve $x^2 = 2\cos y$ at the point $\left(1, \dfrac{\pi}{3}\right)$.

4   Find the equation of the normal to the curve $\ln(2xy) + y^2 = 1$ at the point $\left(\frac{1}{2}, 1\right)$.

5   The diagram shows the graph of the ellipse $4x^2 + y^2 = 8$.
    Find the coordinates of the points on the ellipse where the gradient is 2.

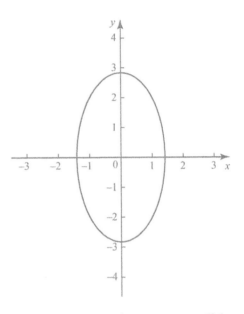

6   Find the coordinates of the point(s) on the curve $x^2 + 6y^2 - 2x + 8y = 39$ where the tangent to the curve is parallel to the x-axis.

7 The equation of a curve is $5x^2 - 2xy + 3y^2 - 70 = 0$.

(i) Show that $\dfrac{dy}{dx} = \dfrac{y - 5x}{3y - x}$.

(ii) Find the coordinates of each of the points on the curve where the tangent is parallel to the x-axis.

(iii) Find the coordinates of each of the points on the curve where the tangent is parallel to the y-axis.

8 The diagram shows the curve defined implicitly by $y^2 + y = x^3 + 2x$.

(i) Find the coordinates of the points of intersection of the curve and the line $x = 2$.

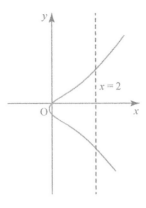

(ii) Find $\dfrac{dy}{dx}$ in terms of x and y and the gradient of the curve at these two points.

OCR MEI Structured Mathematics C3 4753 Paper 01 Q7 May 2005

9 Find the equation of the normal to the curve $x^3 + 4x^2y + y^3 = 6$ at the point $(1, 1)$, giving your answer in the form $ax + by + c = 0$, where a, b and c are integers.

4.6 Parametric equations and parametric differentiation

1 Find $\dfrac{dy}{dx}$ in terms of t or θ for these curves defined parametrically.

(i) $x = 2t$
$y = t^2$

(ii) $x = 2\cos\theta$
$y = 3\sin\theta$

(iii) $x = 3e^t$
$y = t - e^{2t}$

(iv) $x = 2\sec\theta$
$y = 5\tan\theta$

2 A curve is defined parametrically by $x = \dfrac{t}{1-t}$, $y = \dfrac{t^2}{1+t}$.

(i) Find $\dfrac{dy}{dx}$.

(ii) Hence find the coordinates of the stationary points on the curve.

3 The parametric equations of a curve are

$$x = 3e^{4t}, \ y = 6te^{2t}$$

(i) Show that $\dfrac{dy}{dx} = \dfrac{1+2t}{2e^{2t}}$.

(ii) Find the equation of the tangent to the curve at the point where $t = 0$.

4   The graph shows the curve given by the parametric equations

$$x = 2\sin\theta + \cos\theta$$
$$y = \sin\theta + 2\cos\theta$$

(i) Find the equation of the tangent to the curve at the point where $\theta = \dfrac{\pi}{2}$.

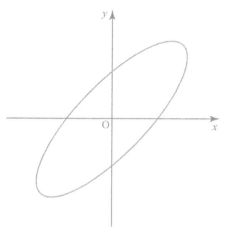

(ii) (a) Show that for any point of the curve, $x^2 + y^2 = 5 + 4\sin 2\theta$.

(b) Hence find the greatest and least distances of a point on the curve from the origin.

5 The parametric equations of a curve are $x = 2t - \ln 2t$, $y = t^2 - \ln t^2$, $t > 0$.

(i) Find $\dfrac{dy}{dx}$ and hence find the exact coordinates of the stationary point on the curve.

(ii) Find the coordinates of the point where $\dfrac{dy}{dx} = 2$.

6 The parametric equations of a curve are $x = 2\theta + \sin 2\theta$, $y = 4 \sin \theta$, and part of its graph is shown below.

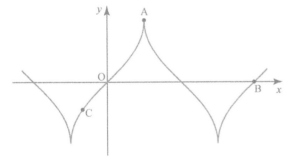

(i) Find the value of θ at A and the value of θ at B.

(ii) Show that $\dfrac{\mathrm{d}y}{\mathrm{d}x} = \sec\theta$.

(iii) At the point C on the curve, the gradient is 2. Find the coordinates of C, giving your answer in an exact form.

7 In a theme park ride, a capsule C moves in a vertical plane as shown in the diagram. With respect to the axes shown, the path of C is modelled by the parametric equations $x = 10\cos\theta + 5\cos 2\theta$, $y = 10\sin\theta + 5\sin 2\theta$, where x and y are in metres.

(i) Show that $\dfrac{\mathrm{d}y}{\mathrm{d}x} = -\dfrac{\cos\theta + \cos 2\theta}{\sin\theta + \sin 2\theta}$.

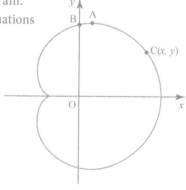

(ii) Hence find the exact coordinates of the highest point A on the path of C.

(iii) Express $x^2 + y^2$ in terms of θ. Hence show that $x^2 + y^2 = 125 + 100\cos\theta$.

(iv) Using this result, or otherwise, find the greatest and least distances of C from O.

OCR MEI Applications of Advanced Mathematics C4 4754(A) Paper 01 Q8 June 2007

Further practice

1   The diagram shows the curve $y = e^{-\frac{1}{4}x}\sqrt{3 + x^2}$.

Find the x-coordinates of the stationary points on the curve.

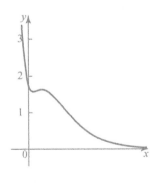

2   The curve shown is $y = \dfrac{(\ln x)^2}{x}$.

Find $\dfrac{dy}{dx}$ and hence find the exact coordinates of the maximum point M.

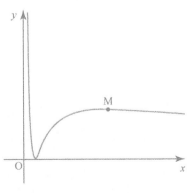

3 The equation of a curve is $y = \sin 2x + x$. Find the coordinates of the stationary points on the curve for $0 \leqslant x \leqslant \pi$, and determine the nature of these stationary points.

4 The equation of a curve is $x^2 + y^2 - xy - 48 = 0$.

   (i) Show that $\dfrac{dy}{dx} = \dfrac{2x - y}{x - 2y}$.

   (ii) Find the coordinates of the points on the curve where the tangent is parallel to the x-axis.

   (iii) Find the coordinates of the points on the curve where the tangent is parallel to the y-axis.

5 The equation of a curve is $2x^2 + xy + y^2 = 14$. Show that there are two stationary points on the curve and find their coordinates.

6 The parametric equations of a curve are $x = \dfrac{2t}{3t + 4}$, $y = 3\ln(3t + 4)$.

   (i) Express $\dfrac{dy}{dx}$ in terms of t, simplifying your answer.

   (ii) Find the gradient of the curve at the point for which $x = 2$.

Past exam questions

1 The equation of a curve is $x^2 + 2xy - y^2 + 8 = 0$.

   (i) Show that the tangent to the curve at the point $(-2, 2)$ is parallel to the x-axis. [4]

   (ii) Find the equation of the tangent to the curve at the other point on the curve for which $x = -2$, giving your answer in the form $y = mx + c$. [5]

   Cambridge International AS & A Level Mathematics 9709 Paper 23 Q8 November 2010

2 The parametric equations of a curve are

   $$x = \ln(\tan t), \quad y = \sin^2 t,$$

   where $0 < t < \frac{1}{2}\pi$.

   (i) Express $\dfrac{dy}{dx}$ in terms of t. [4]

   (ii) Find the equation of the tangent to the curve at the point where $x = 0$. [3]

   Cambridge International AS & A Level Mathematics 9709 Paper 32 Q5 June 2011

3 The parametric equations of a curve are

   $$x = 1 + 2\sin^2\theta, \qquad y = 4\tan\theta.$$

   (i) Show that $\dfrac{dy}{dx} = \dfrac{1}{\sin\theta \, \cos^3\theta}$. [3]

   (ii) Find the equation of the tangent to the curve at the point where $\theta = \frac{1}{4}\pi$, giving your answer in the form $y = mx + c$. [4]

   Cambridge International AS & A Level Mathematics 9709 Paper 22 Q6 November 2011

4 The equation of a curve is $y = \dfrac{e^{2x}}{1 + e^{2x}}$. Show that the gradient of the curve at the point for which $x = \ln 3$ is $\dfrac{9}{50}$. [4]

   Cambridge International AS & A Level Mathematics 9709 Paper 33 Q2 November 2011

5 The equation of a curve is $y = \dfrac{\sin x}{1 + \cos x}$, for $-\pi < x < \pi$.

Show that the gradient of the curve is positive for all x in the given interval. [4]

Cambridge International AS & A Level Mathematics 9709 Paper 33 Q2 November 2016

6 The parametric equations of a curve are $x = t^2 + 1$, $y = 4t + \ln(2t - 1)$.

(i) Express $\dfrac{\mathrm{d}y}{\mathrm{d}x}$ in terms of t. [3]

(ii) Find the equation of the normal to the curve at the point where $t = 1$. Give your answer in the form $ax + by + c = 0$. [3]

Cambridge International AS & A Level Mathematics 9709 Paper 32 Q4 June 2017

▶ STRETCH AND CHALLENGE

1 The path traced out by a point on the circumference of a circle of radius r as the circle rolls along a straight line is called a *cycloid*.

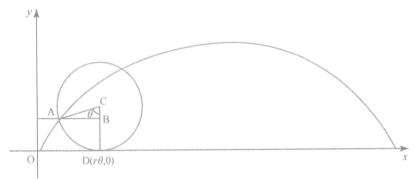

A is a point on the circumference that starts at the origin O. The diagram shows the circle after rotating through θ radians.

(i) Show that the parametric equations of a cycloid are given by

$$x = r(\theta - \sin\theta), \ y = r(1 - \cos\theta) \qquad 0 \leqslant \theta \leqslant 2\pi.$$

(ii) Show that $\dfrac{\mathrm{d}y}{\mathrm{d}x} = \cot\left(\dfrac{\theta}{2}\right)$.

(iii) Find the equation of the tangent to the curve at the point where $y = \frac{1}{2}r$.

(iv) The speed of the point is given by $s = \sqrt{\left(\dfrac{\mathrm{d}x}{\mathrm{d}\theta}\right)^2 + \left(\dfrac{\mathrm{d}y}{\mathrm{d}\theta}\right)^2}$.

Find and simplify an expression for the speed of the point.

(v) Given that acceleration, a, is given by $a = \dfrac{\mathrm{d}s}{\mathrm{d}\theta}$, find and simplify an expression for the acceleration of the point.

2 The parametric equations of a curve are $x = t^3$, $y = t^2$.

(i) Show that the equation of the tangent at the point P where $t = p$ is $3py - 2x = p^3$.

(ii) Given that this tangent passes through the point $(-10, 7)$, find the coordinates of each of the three possible positions of P.

3 The cross-section of a section of gutter for a house is shown.
The sides and base of the gutter are each 10 cm long as shown.

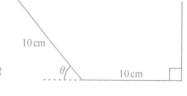

(i) Show that the area, A, of the cross-section is given by
$A = 100\sin\theta + 25\sin 2\theta$.

(ii) Find the maximum area of the cross-section of the gutter and show that this gives a maximum.

4 A ladder is being manoeuvred around a corner in a house.
One hallway is 2 m wide, the other hallway is 3 m wide.

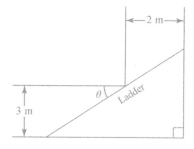

(i) Find an expression for the length of the ladder in terms of θ.

(ii) Hence find the maximum length the ladder can be so it will fit around the corner.

5 Find the cartesian equation of the curve defined parametrically by $x = t + \dfrac{1}{t}$, $y = t - \dfrac{1}{t}$.

6 When a try is scored in a rugby game, the kicker must attempt a conversion from a point directly back from where the try was scored.
The distance between the posts is 5.6 m and a try is scored 30 m to the right of the right-hand post.
Find the value of x (how far back the kicker should go) so that the angle θ between the posts is a maximum.

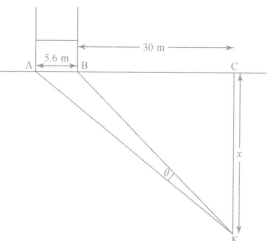

7 The nautilus (shown right) is a marine creature that lives around coral reefs.
The mathematical model of a nautilus shell is an *equiangular spiral*.
Equiangular spirals have equations of the form $r = Ae^{k\theta}$, where k is a constant.
At every point P, the tangent to the curve makes the same angle, α, with the line OP from the point P to the origin (or pole), O.
The size of the angle α depends upon the number k in this mathematical model where $r = Ae^{k\theta}$.

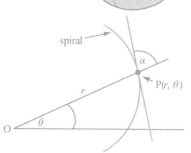

(i) Using the parametric equations for the cartesian coordinates (x, y) of the point P in terms of θ, find $\dfrac{dy}{dx}$.

(ii) Hence, or otherwise, find the value of α in terms of k for this model.

NZQA Scholarship Calculus Q5 2004

5 Integration

5.1 Integrals involving the exponential function

1 Find the following integrals.

(i) $\int e^{2x} dx$

(ii) $\int e^{1-3x} dx$

(iii) $\int \dfrac{4}{e^{2x+3}} dx$

(iv) $\int \dfrac{e^{4x}+1}{e^x} dx$

(v) $\int 9e^{-\frac{1}{3}x} dx$

(vi) $\int (e^x - 1)^2 dx$

2 Find the following definite integral, giving your answer exactly: $\displaystyle\int_1^2 3e^{4x} dx$.

3 Show that $\displaystyle\int_{1}^{\infty} \frac{e^x + 2}{e^{2x}}\,dx = \frac{e+1}{e^2}$.

4 Find $\displaystyle\int \frac{(e^x + 1)^2}{e^{2x}}\,dx$.

5 A curve is such that $\dfrac{dy}{dx} = e^{\frac{1}{3}x} - 3e^{-x}$. The point $(0, -2)$ lies on the curve.

(i) Find the equation of the curve.

(ii) Find the x-coordinate of the stationary point on the curve and determine whether it is a maximum or minimum point.

6   Given that $\int_0^{\ln 4}\left(ke^{3x}+(k-2)e^{-\frac{1}{2}x}\right)dx=185$, find the value of the constant k.

OCR Core Mathematics 3 4723 Paper 01 Q6 January 2010

5.2 Integrals involving the natural logarithm function

1   Find the following integrals.

(i) $\int\dfrac{1}{2x}\,dx$

(ii) $\int\dfrac{2}{x}\,dx$

(iii) $\int\dfrac{3}{2x+1}\,dx$

(iv) $\int\left(\dfrac{1}{3x}+\dfrac{1}{1-2x}\right)dx$

2   Find the following definite integral, giving your answer exactly: $\int_2^4\dfrac{2}{x+1}\,dx$.

3 Show that $\displaystyle\int_{-1}^{1} \frac{9}{1-3x}\,dx = \ln 8$.

4 Find $\displaystyle\int \frac{x-1}{x^2-1}\,dx$.

5 Find $\displaystyle\int \frac{3x^2-7x}{3x+2}\,dx$.

6 The area between the curve $y = \dfrac{3}{3x-1}$ and the x-axis between $x = k$ and $x = 1$ is exactly 1. Find the exact value of k.

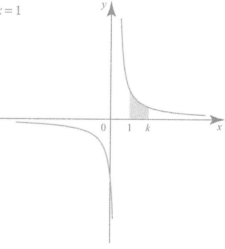

5.3 Integrals involving trigonometrical functions

1 Find the following integrals.

   (i) $\int \sin 4x \, dx$

   (ii) $\int \cos(3x-1) \, dx$

   (iii) $\int \left(4\cos\frac{1}{2}x + 1 \right) dx$

   (iv) $\int (\cos 2x - \sec^2 3x + 4) \, dx$

2 Find the exact values of these definite integrals.

   (i) $\int_0^{\frac{\pi}{4}} \sin 2x \, dx$

   (ii) $\int_{\frac{\pi}{3}}^{\frac{\pi}{2}} \sec^2 \frac{1}{2}x \, dx$

3 Find the area enclosed between the curve $y = \cos x$, the line $y = \frac{1}{2}$ and the y-axis for $0 \leqslant x \leqslant \frac{\pi}{2}$.

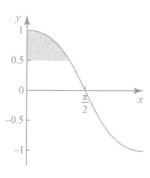

4 Find the following integrals by using a trigonometric identity.

(i) $\displaystyle\int \sin^2 x \, dx$

(ii) $\displaystyle\int 4 \sin x \cos x \, dx$

(iii) $\displaystyle\int \sin^4 x \, dx$

(iv) $\displaystyle\int (\sin 4x \cos x - \cos 4x \sin x) \, dx$

5 Show that $\displaystyle\int_0^{\frac{\pi}{4}} \sqrt{1 + \cos 4x} \, dx = \frac{\sqrt{2}}{2}$.

6 The diagram shows the curves $y = \sin 2x$ and $y = \cos x$ for $-\frac{\pi}{2} \leqslant x \leqslant \frac{\pi}{2}$.
Find the shaded area.

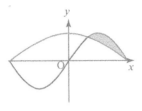

7 The diagram shows the curve $y = 2\sin x$ for $0 \leqslant x \leqslant \pi$ and the line $y = 1$.

(i) Find the area enclosed by the curve and the line.

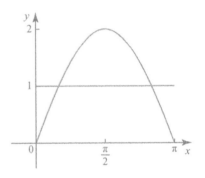

(ii) The region enclosed by the curve and the line is rotated 360° around the x-axis.
Find the volume of the solid generated.

8 (i) Show that $\cos 3x \equiv 4\cos^3 x - 3\cos x$.

(ii) Hence find the exact value of $\int_{\frac{\pi}{3}}^{\frac{\pi}{2}} \cos^3 x \, dx$.

9 (i) Given that $5\cos x - 3\sin x = A(\cos x + \sin x) + B(\cos x - \sin x)$ for all values of x, find the values of the constants A and B.

(ii) Hence find the exact value of $\int_0^{\frac{\pi}{2}} \frac{5\cos x - 3\sin x}{\cos x + \sin x} \, dx$.

10 By expressing $\cos 2x$ in terms of $\cos x$, find the exact value of $\int_{\frac{\pi}{6}}^{\frac{\pi}{3}} \frac{\cos 2x}{\cos^2 x}\, dx$.

5.4 Numerical integration (Pure Mathematics 2 only)

1   The values of x and $f(x)$ are given in the table below for $0 \leqslant x \leqslant 2$.

x	0	0.5	1	1.5	2
$f(x)$	2.1	0.8	1.5	1.9	3.8

Use the trapezium rule with 4 strips to find an estimate for $\int_0^2 f(x)\, dx$.

2   The function $f(x)$ is defined by $f(x) = \dfrac{1}{x^3 + 1}$.

   (i) Complete the table of values for $f(x)$.

x	0	0.25	0.5	0.75	1
$f(x)$	1				0.5

(ii) Use the trapezium rule with 4 strips to estimate the area between the curve and the x-axis between $x = 0$ and $x = 1$.

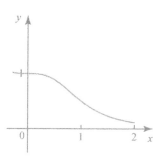

(iii) Using the diagram, determine whether the answer from part (ii) is an overestimate or underestimate of the true area under the curve.

3   The diagram shows the cross-section of a certain section of river. The depth of the river is measured as shown. The measurements are made 0.5 m apart.

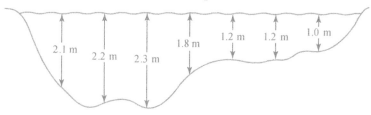

(i) Use the trapezium rule to find an estimate of the area of the cross-section.

(ii) If the river is flowing at a constant 3 km/h, find the volume of water in m³ passing this spot of the river every minute.

4   The diagram shows part of the curve $y = \sqrt{1+x^3}$.

   (i) Use the trapezium rule with 4 strips to estimate $\int_0^2 \sqrt{1+x^3}\, dx$, giving your answer correct to 3 significant figures.

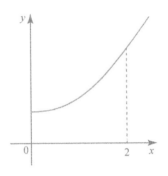

(ii) Chris and Dave each estimate the value of this integral using the trapezium rule with 8 strips. Chris gets a result of 3.25, and Dave gets 3.30. One of these results is correct.
Without performing the calculation, state with a reason which is correct.

OCR MEI Applications of Advanced Mathematics C4 4754(A) Paper 01 Q2 January 2007

Further practice

1 (i) Show that $\displaystyle\int_{-2}^{2} \frac{4}{2x+1}\,dx = \ln\frac{25}{9}$.

   (ii) Find the value of k $(k < 0)$ such that $\displaystyle\int_{k}^{2} \frac{4}{2x+1}\,dx = 0$.

2 (i) Show that $(2\cos x + \sin x)^2$ can be written in the form $a\sin 2x + b\cos 2x + c$, stating the values of a, b and c.

   (ii) Hence find the exact value of $\displaystyle\int_{0}^{\frac{\pi}{2}} (2\cos x + \sin x)^2\,dx$.

3 Show that $\tan^2 x + \sin^2 x = \sec^2 x - \frac{1}{2}\cos 2x - \frac{1}{2}$ and hence find the exact value of

   $\displaystyle\int_{0}^{\frac{\pi}{6}} \left(\tan^2 x + \sin^2 x\right)\,dx$

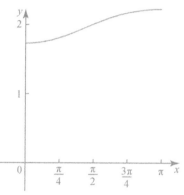

4 The diagram shows a part of the curve $y = \sqrt{4 - \cos x}$ for $0 \leqslant x \leqslant \pi$.

   (i) Use the trapezium rule with two intervals to estimate the value of
   $\displaystyle\int_{0}^{\pi} \sqrt{4 - \cos x}\,dx$ correct to 3 significant figures.

   (ii) Explain, with reference to the diagram, why the trapezium rule may be expected to give a good approximation to the true value of the integral in this case.

5 (i) Differentiate $e^x(\sin 2x - 2\cos 2x)$, simplifying your answer.

   (ii) Hence find the exact value of $\displaystyle\int_{0}^{\frac{1}{4}\pi} e^x \sin 2x\,dx$.

6 (i) Express $\cos\theta + \sqrt{3}\sin\theta$ in the form $R\cos(\theta - \alpha)$ where $R > 0$ and α is acute. Express α in terms of π.

   (ii) Hence show that $\displaystyle\int_{0}^{\frac{\pi}{3}} \frac{1}{(\cos\theta + \sqrt{3}\sin\theta)^2}\,d\theta = \frac{\sqrt{3}}{4}$.

Past exam questions

1 (i) By differentiating $\frac{\cos x}{\sin x}$, show that if $y = \cot x$ then $\frac{dy}{dx} = -\mathrm{cosec}^2 x$. [3]

  (ii) By expressing $\cot^2 x$ in terms of $\mathrm{cosec}^2 x$ and using the result of part (i), show that

$$\int_{\frac{1}{4}\pi}^{\frac{1}{2}\pi} \cot^2 x \, dx = 1 - \frac{1}{4}\pi.$$ [4]

  (iii) Express $\cos 2x$ in terms of $\sin^2 x$ and hence show that $\frac{1}{1 - \cos 2x}$ can be expressed as $\frac{1}{2}\mathrm{cosec}^2 x$.

Hence, using the result of part (i), find $\int \frac{1}{1 - \cos 2x} \, dx$. [3]

Cambridge International AS & A Level Mathematics 9709 Paper 21 Q8 June 2010

2 (i) Using the expansions of $\cos(3x - x)$ and $\cos(3x + x)$, prove that $\frac{1}{2}(\cos 2x - \cos 4x) \equiv \sin 3x \sin x$. [3]

  (ii) Hence show that $\int_{\frac{1}{6}\pi}^{\frac{1}{3}\pi} \sin 3x \sin x \, dx = \frac{1}{8}\sqrt{3}$. [3]

Cambridge International AS & A Level Mathematics 9709 Paper 31 Q4 June 2010

3 The diagram shows the curve $y = \sqrt{(1 + x^3)}$. Region A is bounded by the curve and the lines $x = 0$, $x = 2$ and $y = 0$. Region B is bounded by the curve and the lines $x = 0$ and $y = 3$.

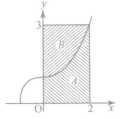

  (i) Use the trapezium rule with two intervals to find an approximation to the area of region A. Give your answer correct to 2 decimal places. [3]

  (ii) Deduce an approximation to the area of region B and explain why this approximation underestimates the true area of region B. [2]

Cambridge International AS & A Level Mathematics 9709 Paper 22 Q2 June 2011

4 Find the exact value of the positive constant k for which

$$\int_0^k e^{4x} \, dx = \int_0^{2k} e^x \, dx.$$ [6]

Cambridge International AS & A Level Mathematics 9709 Paper 22 Q4 November 2011

5 (a) Find $\int \frac{4 + e^x}{2e^{2x}} \, dx$. [3]

  (b) Without using a calculator, find $\int_2^{10} \frac{1}{2x + 5} \, dx$, giving your answer in the form $\ln k$. [4]

  (c) The diagram shows the curve $y = \log_{10}(x + 2)$ for $0 \leqslant x \leqslant 6$.

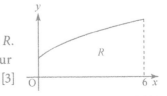

The region bounded by the curve and the lines $x = 0$, $x = 6$ and $y = 0$ is denoted by R. Use the trapezium rule with 2 strips to find an estimate of the area of R, giving your answer correct to 1 decimal place. [3]

Cambridge International AS & A Level Mathematics 9709 Paper 22 Q6 June 2016

6 (i) Find $\int (2\cos\theta - 3)(\cos\theta + 1) \, d\theta$. [4]

  (ii) (a) Find $\int \left(\frac{4}{2x + 1} + \frac{1}{2x} \right) dx$. [2]

    (b) Hence find $\int_1^4 \left(\frac{4}{2x + 1} + \frac{1}{2x} \right) dx$, giving your answer in the form $\ln k$. [3]

Cambridge International AS & A Level Mathematics 9709 Paper 22 Q7 June 2017

STRETCH AND CHALLENGE

. . . .

1   The length of a curve between $x = b$ and $x = a$ is given by the formula $L = \int_a^b \sqrt{1 + [f'(x)]^2}\,dx$.

Show that the length of the curve $y = \dfrac{x^3}{6} + \dfrac{1}{2x}$ between $x = 2$ and $x = 1$ is $\dfrac{17}{12}$.

2   Show that $\int_0^{\frac{\pi}{4}} \sin^2 x \cos^2 x\,dx = \dfrac{\pi}{32}$.

3   A curve in polar form is given in terms of its distance from the origin (r) and angle made with the positive x-axis (θ).

So a point with $r = 2$ and $\theta = \dfrac{\pi}{6}$ is given in polar form by $\left[2, \dfrac{\pi}{6}\right]$.

Some polar curves are shown below.

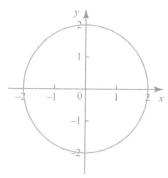

$r = 2$

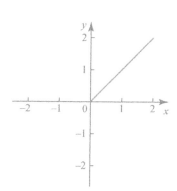

$\theta = \dfrac{\pi}{4}$

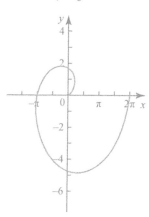

$r = \theta$

(i)   The diagram shows the polar curves $r = k\theta$, $\theta = m$ and $r = t$ (where k, m and t are constants).

If point P has polar coordinates $\left[\dfrac{3\pi}{2}, \dfrac{3\pi}{4}\right]$, determine the exact values of k, m and t.

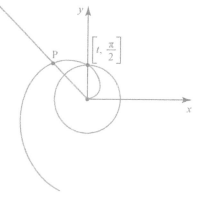

(ii)   If $r = p(\theta)$, the length of a polar curve is given by $L = \int_0^{2\pi} \sqrt{[r(\theta)]^2 + [r'(\theta)]^2}\,d\theta$.

Find the length of the cardioid $r = 1 + \cos\theta$.

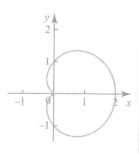

(iii) The parametric equations of a cycloid are given by

$$x = r(\theta - \sin\theta), \; y = r(1 - \cos\theta), \; 0 \leqslant \theta \leqslant 2\pi$$

Find the length of the cycloid which is given by

$$L = \int_0^{2\pi} \sqrt{\left(\frac{\mathrm{d}x}{\mathrm{d}\theta}\right)^2 + \left(\frac{\mathrm{d}y}{\mathrm{d}\theta}\right)^2} \, \mathrm{d}\theta$$

4   The Moeraki boulders are natural stone spheres sunk into the sand of Moeraki Beach between Oamaru and Dunedin.

The diagram shows a cross-section of the side view of a partially submerged Moeraki boulder.

The angle between the surface of the water and a tangent plane to the boulder is ϕ, as shown.

Find the **proportion** of the volume of the boulder which is below water level.

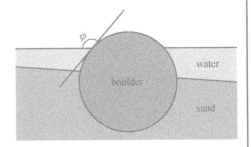

NZQA Scholarship Calculus Q3c 2011

6 Numerical solution of equations

6.1 Interval estimation – change-of-sign methods

1 The diagram shows the graphs of $y = e^{x-3}$ and $y = x^3$.

To find where the two curves intersect, solve the equation $e^{x-3} = x^3$.

(i) Rearrange the equation so it is in the form $f(x) = 0$.

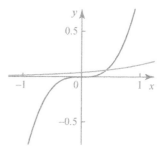

(ii) Show that the root of $f(x) = 0$ lies between 0 and 1.

(iii) Show that the root of $f(x) = 0$ lies between 0 and 0.5.

(iv) Find the two values of x, correct to 1 decimal place, between which the root lies.

2   For the curve $g(x) = \frac{x+4}{x-1}$, $g(0) = -4$ and $g(2) = 6$.

As there is a change in sign between $x = 0$ and $x = 2$, there must be a root of the equation $g(x) = 0$ between $x = 0$ and $x = 2$.

State, with a reason, why this statement is false in this case.

3   (i)   By drawing a sketch, show that the equation $4 - x = \ln x$ has only one root.

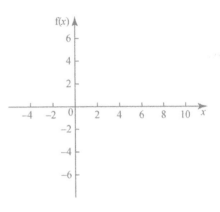

(ii)   Rearrange the equation $4 - x = \ln x$ so it is in the form $f(x) = 0$.

(iii)   Find the integers a and b such that $f(a) > 0$ and $f(b) < 0$ or $f(a) < 0$ and $f(b) > 0$.
Hence state the integer bounds between which the root of $f(x) = 0$ lies.

(iv) Verify by calculation that this root lies between 2.9 and 3.0.

4 (i) Show that the equation $x^2 = 3^x$ has a root α in the interval $-0.7 < \alpha < -0.6$.

(ii) Sketch the graphs of $y = x^2$ and $y = 3^x$ to verify there is just one root of the equation $x^2 = 3^x$.

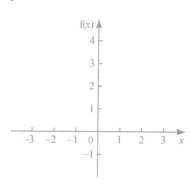

6.2 Fixed-point iteration and problems with the method

1   The equation $x^3 - 2x + 3 = 0$ has one root.

   (i) Sketch the graphs of $y = x^3$ and $y = 2x - 3$ on the axes provided.

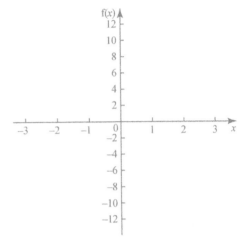

   (ii) Use your diagram to find the integers between which the root lies.

   (iii) Use the iterative formula $x_{n+1} = \sqrt[3]{2x_n - 3}$ to find the root correct to 4 decimal places.

   (iv) Suggest two other arrangements of the form $x_{n+1} = f(x_n)$ that could be used to find the root.

2 The equation $4 - x = \ln x$ has a root α where $2.9 < \alpha < 3.0$.

Use the iterative formula $x_{n+1} = 4 - \ln x_n$ with initial value $x_0 = 2.9$ to find the value of α correct to 3 decimal places. Give the result of each iteration to 4 decimal places.

3 **(i)** Show that the equation $x^3 - x^2 = 15$ has a root between $x = 2$ and $x = 3$.

(ii) Use the iterative formula $x_{n+1} = \sqrt[3]{15 + x_n^2}$ with $x_0 = 2.5$ to find the root correct to 3 decimal places.

4 (i) Show that if the iterative formula $x_{n+1} = \sqrt{\dfrac{3 - x_n}{x_n}}$ converges to the value α, then α will be a root of the equation $x^3 + x - 3 = 0$.

(ii) Use the iterative formula with $x_0 = 1.5$ to find the value of α correct to 2 decimal places.

5 The diagram shows a shaded segment of a circle centre O and radius r.

(i) Show that the area, S, of the segment is given by $S = \frac{1}{2}r^2(\theta - \sin\theta)$.

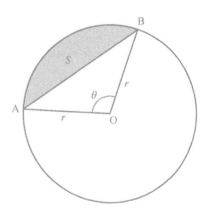

(ii) The chord AB divides the area of the circle in the ratio $1:5$.

Show that θ satisfies $\theta = \frac{1}{3}\pi + \sin\theta$.

(iii) Use the iterative formula $\theta_{n+1} = \frac{1}{3}\pi + \sin\theta_n$ with $\theta_1 = 1$ to find θ correct to 2 decimal places.

6 The equation $x^2 = 3^x$ has a root α in the interval $-0.7 < \alpha < -0.6$.

Show that the following three possible arrangements $x_{n+1} = f(x_n)$ fail to converge to the root.

(i) $x_{n+1} = \sqrt{3^{x_n}}$

(ii) $x_{n+1} = \log_3(x_n^{\ 2})$

(iii) $x_{n+1} = \dfrac{3^{x_n}}{x_n}$

7 The sequence defined by

$$x_1 = 3, \qquad x_{n+1} = \sqrt[3]{31 - \frac{5}{2}x_n}$$

converges to the number α.

(i) Find the value of α correct to 3 decimal places, showing the result of each iteration.

(ii) Find an equation of the form $ax^3 + bx + c = 0$, where a, b and c are integers, which has α as a root.

OCR Core Mathematics 3 4723 Paper 01 Q2 January 2008

8 (i) It is given that k is a positive constant. By sketching the graphs of $y = 14 - x^2$ and $y = k \ln x$ on a single diagram, show that the equation $14 - x^2 = k \ln x$ has exactly one real root.

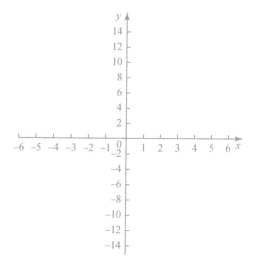

(ii) The real root of the equation $14 - x^2 = 3 \ln x$ is denoted by α.

   (a) Find by calculation the pair of consecutive integers between which α lies.

   (b) Use the iterative formula $x_{n+1} = \sqrt{14 - 3 \ln x_n}$, with a suitable starting value, to find α. Show the result of each iteration, and give α correct to 2 decimal places.

OCR Core Mathematics 3 4723 Paper 01 Q5 June 2012

Further practice

1 The line $y = x$ intersects the curve $y = \sqrt{4 - \cos x}$ at the point M. Use the iterative formula

$$x_{n+1} = \sqrt{4 - \cos x_n}$$

with $x_0 = 5$ to determine the x-coordinate of M correct to 2 decimal places. Give the result of each iteration to 4 decimal places.

2 A curve is given by $y = e^{-\frac{1}{4}x} \sqrt{3 + x^2}$.

   (i) The sequence of values given by the iterative formula

   $$x_{n+1} = 2\ln(48 + 16x_n^2)$$

   with initial value $x_0 = 14$ converges to a certain value α. State an equation satisfied by α and hence show that α is the x-coordinate of a point on the curve where $y = 0.25$.

   (ii) Use the iterative formula to calculate the value of α to 2 decimal places. Give the result of each iteration to 4 decimal places.

3 The diagram shows the curve $y = x^4 - 4x^3 + 4x^2 + 2x - 7$, which crosses the x-axis at the points $(\alpha, 0)$ and $(\beta, 0)$ where $\alpha < \beta$. It is given that α is an integer.

   (i) Find the value of α.

   (ii) Show that β satisfies the equation $x = \sqrt[3]{5x^2 - 9x + 7}$.

   (iii) Use an iteration process based on the equation in part (ii) to find the value of β correct to 2 decimal places. Show the result of each iteration to 4 decimal places.

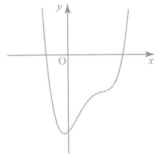

4 The diagram shows the curve $y = \frac{\cos 2x}{1 - x}$. The x-coordinate of the maximum point M is denoted by α.

   (i) Find $\frac{dy}{dx}$ and show that α satisfies the equation $\tan 2x = \frac{1}{2 - 2x}$.

   (ii) Show by calculation that α lies between 0.3 and 0.4.

   (iii) Use the iterative formula $x_{n+1} = \frac{1}{2}\tan^{-1}\left(\frac{1}{2 - 2x_n}\right)$ to find the value of α correct to 3 decimal places. Give the result of each iteration to 5 decimal places.

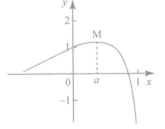

5 The diagram shows a circle radius r and centre O with the radius OB extended to meet the tangent to the circle at A at the point C. The shaded area is the same as the area of the sector OAB.

   (i) Show that θ satisfies the equation $2\theta = \tan\theta$.

   (ii) This equation has one root in the interval $0 < \theta < \frac{\pi}{2}$. Use the iterative formula $\theta_{n+1} = \tan^{-1}(2\theta_n)$ to find the root correct to 2 decimal places. Give the result of each iteration correct to 4 decimal places.

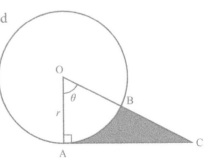

Past exam questions

1 The diagram shows a semicircle ACB with centre O and radius r. The tangent at C meets AB produced at T. The angle BOC is x radians. The area of the shaded region is equal to the area of the semicircle.

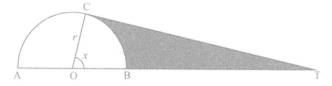

(i) Show that x satisfies the equation $\tan x = x + \pi$. [3]

(ii) Use the iterative formula $x_{n+1} = \tan^{-1}(x_n + \pi)$ to determine x correct to 2 decimal places. Give the result of each iteration to 4 decimal places. [3]

Cambridge International AS & A Level Mathematics 9709 Paper 32 Q4 June 2011

2 (i) By sketching a suitable pair of graphs, show that the equation $\frac{1}{x} = \sin x$, where x is in radians, has only one root for $0 < x \leqslant \frac{1}{2}\pi$. [2]

(ii) Verify by calculation that this root lies between $x = 1.1$ and $x = 1.2$. [2]

(iii) Use the iterative formula $x_{n+1} = \frac{1}{\sin x_n}$ to determine this root correct to 2 decimal places. Give the result of each iteration to 4 decimal places. [3]

Cambridge International AS & A Level Mathematics 9709 Paper 22 Q5 November 2011

3 It is given that $\int_0^a (3e^{3x} + 5e^x)\,dx = 100$, where a is a positive constant.

(i) Show that $a = \frac{1}{3}\ln(106 - 5e^a)$. [5]

(ii) Use an iterative formula based on the equation in part (i) to find the value of a correct to 3 decimal places. Give the result of each iteration to 5 decimal places. [3]

Cambridge International AS & A Level Mathematics 9709 Paper 23 Q5 November 2015

4 The sequence of values given by the iterative formula

$$x_{n+1} = \frac{4}{x_n^2} + \frac{2x_n}{3},$$

with initial value $x_1 = 2$, converges to α.

(i) Use this iterative formula to find α correct to 3 decimal places. Give the result of each iteration to 5 decimal places. [3]

(ii) State an equation that is satisfied by α, and hence find the exact value of α. [2]

Cambridge International AS & A Level Mathematics 9709 Paper 23 Q1 November 2016

▶ STRETCH AND CHALLENGE

1 The *secant method* is another way to find the root of an equation.

It requires two starting points, x_1 and x_2, but they need not be on opposite sides of the exact solution.

A straight line is drawn through the two points $(x_1, f(x_1))$ and $(x_2, f(x_2))$, and the next estimate is taken as the point at which this line cuts the x-axis.

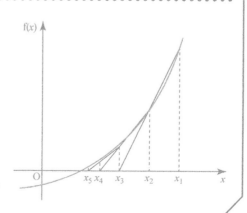

(i) Develop an equation for x_{n+1} in terms of x_n and x_{n-1}.

(ii) Use the secant method to find x_3 for the equation $f(x) = e^{2x} - 3$, given that $x_1 = 1$ and $x_2 = 0$.

Formula sheet

Mensuration

Volume of sphere $= \frac{4}{3}\pi r^3$

Surface area of sphere $= 4\pi r^2$

Volume of cone or pyramid $= \frac{1}{3} \times$ base area $\times$ height

Area of curved surface of cone $= \pi r \times$ slant height

Arc length of circle $= r\theta$ (θ in radians)

Area of sector of circle $= \frac{1}{2}r^2\theta$ (θ in radians)

Algebra

For the quadratic equation $ax^2 + bx + c = 0$:

$$x = \frac{-b \pm \sqrt{b^2 - 4ac}}{2a}$$

Trigonometry

$$\tan\theta \equiv \frac{\sin\theta}{\cos\theta}$$

$$\cos^2\theta + \sin^2\theta \equiv 1, \qquad 1 + \tan^2\theta \equiv \sec^2\theta, \qquad \cot^2\theta + 1 \equiv \mathrm{cosec}^2\theta$$

$$\sin(A \pm B) \equiv \sin A\cos B \pm \cos A\sin B$$

$$\cos(A \pm B) \equiv \cos A\cos B \mp \sin A\sin B$$

$$\tan(A \pm B) \equiv \frac{\tan A \pm \tan B}{1 \mp \tan A\tan B}$$

$$\sin 2A \equiv 2\sin A\cos A$$

$$\cos 2A \equiv \cos^2 A - \sin^2 A \equiv 2\cos^2 A - 1 \equiv 1 - 2\sin^2 A$$

$$\tan 2A \equiv \frac{2\tan A}{1 - \tan^2 A}$$

Principal values:

$$-\frac{1}{2}\pi \leqslant \sin^{-1}x \leqslant \frac{1}{2}\pi, \qquad 0 \leqslant \cos^{-1}x \leqslant \pi, \qquad -\frac{1}{2}\pi < \tan^{-1}x < \frac{1}{2}\pi$$

Differentiation

f(x)	f$'$(x)
x^n	nx^{n-1}
$\ln x$	$\dfrac{1}{x}$
e^x	e^x
$\sin x$	$\cos x$
$\cos x$	$-\sin x$
$\tan x$	$\sec^2 x$
$\sec x$	$\sec x \tan x$
$\operatorname{cosec} x$	$-\operatorname{cosec} x \cot x$
$\cot x$	$-\operatorname{cosec}^2 x$
$\tan^{-1} x$	$\dfrac{1}{1+x^2}$
uv	$v\dfrac{du}{dx} + u\dfrac{dv}{dx}$
$\dfrac{u}{v}$	$\dfrac{v\dfrac{du}{dx} - u\dfrac{dv}{dx}}{v^2}$

If $x = f(t)$ and $y = g(t)$ then $\dfrac{dy}{dx} = \dfrac{dy}{dt} \div \dfrac{dx}{dt}$

Integration

(Arbitrary constants are omitted; a denotes a positive constant.)

f(x)	$\displaystyle\int$ f(x) dx	
x^n	$\dfrac{x^{n+1}}{n+1}$	$(n \neq -1)$
$\dfrac{1}{x}$	$\ln\lvert x \rvert$	
e^x	e^x	
$\sin x$	$-\cos x$	
$\cos x$	$\sin x$	
$\sec^2 x$	$\tan x$	

$$\int \frac{f'(x)}{f(x)} \, dx = \ln\lvert f(x) \rvert$$

Reinforce learning and deepen understanding of the key concepts covered in the latest syllabus; an ideal course companion or homework book for use throughout the course.

» Develop and strengthen skills and knowledge with a wealth of additional exercises that perfectly supplement the Paper 2 content in the Student's Book.

» Build confidence with extra practice for each lesson to ensure that a topic is thoroughly understood before moving on.

» Ensure students know what to expect with hundreds of rigorous practice and exam-style questions.

» Keep track of students' work with ready-to-go write-in exercises.

» Save time with all answers available online at:
www.hoddereducation.com/cambridgeextras.

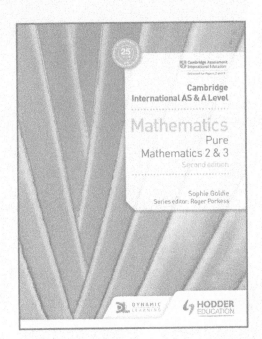

Use with *Cambridge International AS & A Level Mathematics Pure Mathematics 2 & 3 Second edition*

9781510421738

For over 30 years we have been trusted by Cambridge schools around the world to provide quality support for teaching and learning. For this reason we have been selected by Cambridge Assessment International Education as an official publisher of endorsed material for their syllabuses.

www.hoddereducation.com

ISBN 978-1-5104-5843-7

9 781510 458437